Bugs

Cockroaches

by Trudy Becker

FOCUS READERS
PIONEER

www.focusreaders.com

Focus Readers is distributed by North Star Editions:
sales@northstareditions.com | 888-417-0195

Produced for Focus Readers by Red Line Editorial.

Photographs ©: Shutterstock Images, cover, 1, 4, 6, 8, 10, 12, 14, 17, 18, 20 (top), 20 (bottom)

Library of Congress Cataloging-in-Publication Data
Names: Becker, Trudy, author.
Title: Cockroaches / by Trudy Becker.
Description: Lake Elmo, MN : Focus Readers, [2023] | Series: Bugs | Includes index. | Audience: Grades 2-3
Identifiers: LCCN 2022031700 (print) | LCCN 2022031701 (ebook) | ISBN 9781637394496 (hardcover) | ISBN 9781637394861 (paperback) | ISBN 9781637395592 (pdf) | ISBN 9781637395233 (ebook)
Subjects: LCSH: Cockroaches--Juvenile literature.
Classification: LCC QL505.5 .B425 2023 (print) | LCC QL505.5 (ebook) | DDC 595.7/28--dc23/eng/20220713
LC record available at https://lccn.loc.gov/2022031700
LC ebook record available at https://lccn.loc.gov/2022031701

Printed in the United States of America
Mankato, MN
012023

About the Author

Trudy Becker lives in Minneapolis, Minnesota. She likes exploring new places and loves anything involving books.

Table of Contents

Chapter 1

Night Movement

It is nighttime. A cockroach scurries across the ground. Its **antennae** reach out. It finds a juicy piece of fruit. The cockroach has a feast. Then it scurries off again.

Cockroaches live in dark, warm places. In the wild, they stay in caves or under the forest floor. But some cockroaches live inside buildings. They even find places to hide in people's homes.

Fun Fact

Antarctica is the only continent without cockroaches.

Chapter 2

Strong Bodies

Cockroaches have long, oval bodies. They have six strong legs. They have wings, too. But most cockroaches don't fly very often. Some don't fly at all. They run fast instead.

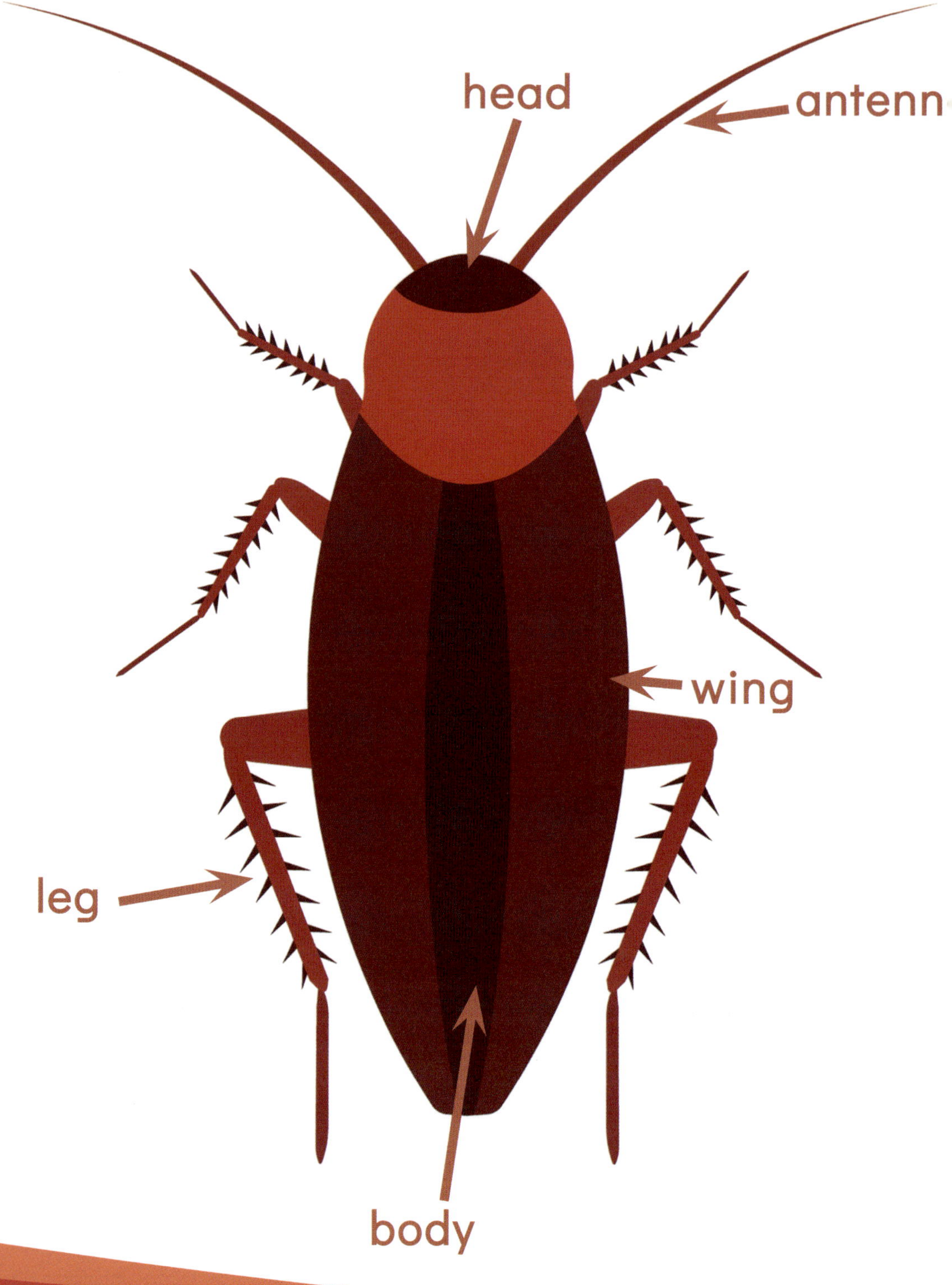
head
antenn
wing
leg
body

Like many insects, cockroaches have antennae. Their antennae are long and thin. Cockroaches use them to sense what is around. Their legs can sense things, too.

Fun Fact

Cockroaches can survive underwater for 40 minutes.

Chapter 3

Hard to Kill

Cockroaches are hard to kill. Their **exoskeletons** are very strong. So, it's hard to crush them. Some animals can eat them, though. Frogs and mice are two of their **predators**.

Cockroaches can go a long time without eating. If they have water, they can last a whole month. That is because cockroaches are **cold-blooded**. They save their energy by resting.

Fun Fact

A cockroach can survive being crushed by 900 times its body weight.

No Head? No Problem!

Sometimes a cockroach loses its head. But it doesn't die right away. It can still live for a whole week. It can't eat or drink. But it can breathe through holes on its body. And its **nerves** help it keep moving.

Chapter 4

Roach Life

Cockroaches are **omnivores**. They eat anything they can find. They eat plants or animals. They eat sweet things. They even eat moldy things.

nymph

egg case

Young cockroaches are called **nymphs**. They eat rotten food. At first, nymphs are tiny. But they can run fast. And they grow into strong adults.

Fun Fact

Cockroaches lay eggs in cases. Each case looks like a small pouch.

FOCUS ON

Cockroaches

Write your answers on a separate piece of paper.

1. Write a sentence that explains the main idea of Chapter 3.
2. Cockroaches eat almost anything they can find. What do you like to eat?
3. What do young cockroaches eat?
    A. frogs
    B. mice
    C. rotten food
4. Why are strong exoskeletons helpful for cockroaches?
    A. They protect cockroaches from danger.
    B. They are heavy, so cockroaches don't blow away.
    C. They help cockroaches run faster.

Answer key on page 24.

Glossary

antennae
Long, thin body parts on an insect's head. The parts are used for sensing.

cold-blooded
Having a body temperature that matches the temperature of the surrounding water or air.

exoskeletons
Hard shells that protect animals' bodies.

nerves
Body parts that sense things and send messages to the brain.

nymphs
Baby insects.

omnivores
Animals that eat both meat and plants to survive.

predators
Animals that hunt other animals for food.

To Learn More

BOOKS

Amstutz, Lisa J. *Fast Facts About Cockroaches.* North Mankato, MN: Capstone Press, 2021.

Perish, Patrick. *Cockroaches.* Minneapolis: Bellwether Media, 2019.

NOTE TO EDUCATORS

Visit **www.focusreaders.com** to find lesson plans, activities, links, and other resources related to this title.

Index

Answer Key: 1. Answers will vary; **2.** Answers will vary; **3.** C; **4.** A